DeltaScience ContentReaders

Our Solar System and Beyond

Contents

Preview the Book 2
What Is in Our Solar System? 3
   About the Solar System. 4
   The Sun 7
   The Planets 8
      The Inner Planets 10
      The Outer Planets 12
   Other Objects in the Solar System 14

Main Idea and Details 16
What Is Beyond Our Solar System? 17
   Other Stars 18
   Galaxies 20
   The Universe 21
   Studying and Exploring Space 22

Glossary 24

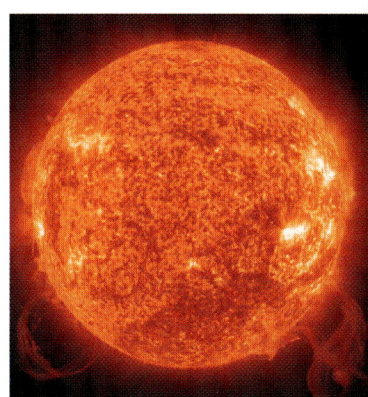

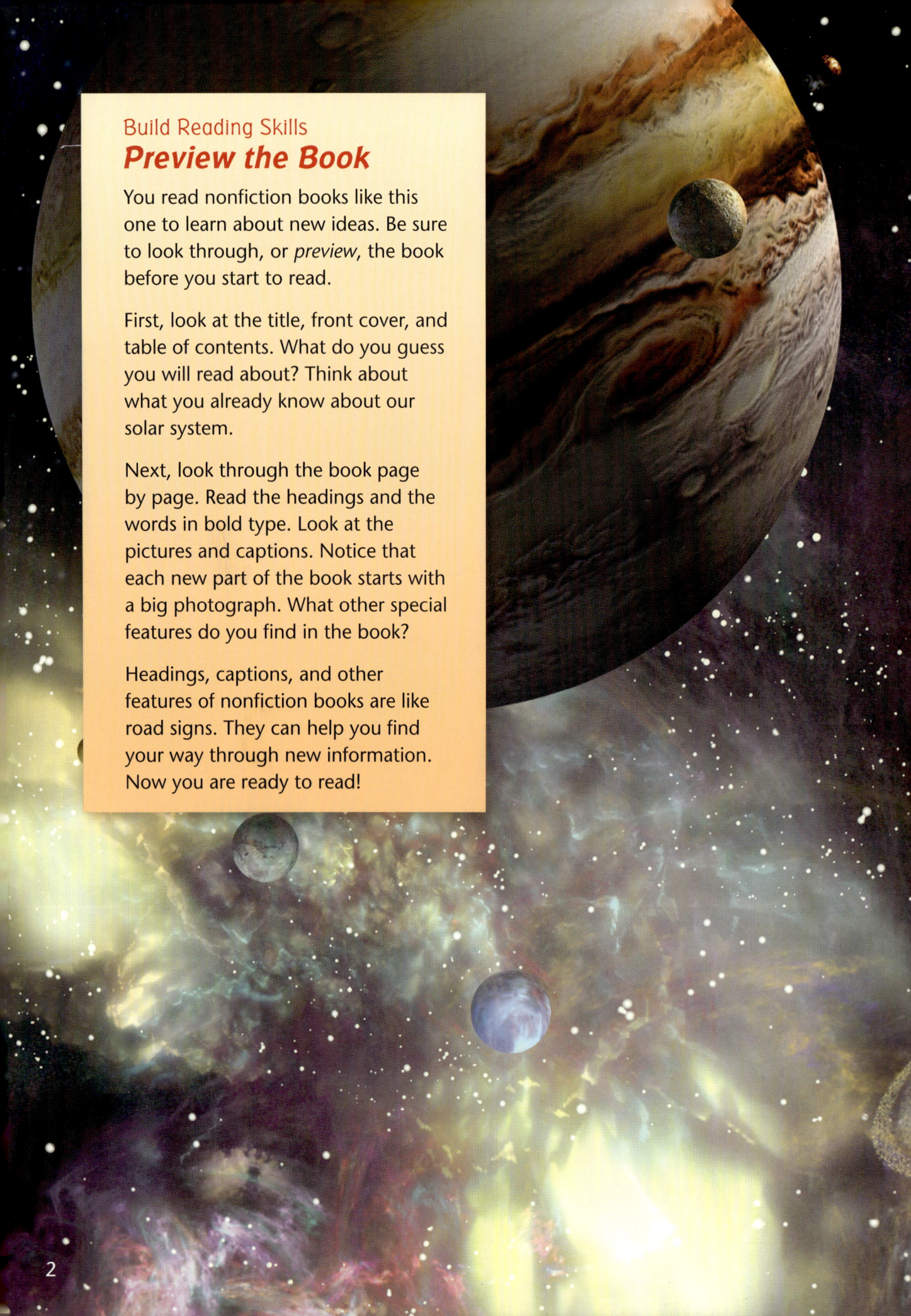

Build Reading Skills
Preview the Book

You read nonfiction books like this one to learn about new ideas. Be sure to look through, or *preview*, the book before you start to read.

First, look at the title, front cover, and table of contents. What do you guess you will read about? Think about what you already know about our solar system.

Next, look through the book page by page. Read the headings and the words in bold type. Look at the pictures and captions. Notice that each new part of the book starts with a big photograph. What other special features do you find in the book?

Headings, captions, and other features of nonfiction books are like road signs. They can help you find your way through new information. Now you are ready to read!

What Is in Our Solar System?

MAKE A CONNECTION
The planets are not close together, the way they are shown in this picture. But they *are* different sizes. Earth is shown as a small blue ball. Can you name any other planets?

FIND OUT ABOUT
- the solar system
- the Sun
- the planets and their moons
- dwarf planets, asteroids, comets, and meteoroids

VOCABULARY
revolve, p. 4
star, p. 4
planet, p. 4
solar system, p. 4
orbit, p. 5
rotate, p. 5
axis, p. 5
Sun, p. 7
moon, p. 8
dwarf planet, p. 14
asteroid, p. 14
comet, p. 15
meteor, p. 15

About the Solar System

Space has been studied for thousands of years. People once believed that Earth always stayed in the same place in space. They thought that the Sun moved around Earth. About 500 years ago, people began to understand that those ideas are incorrect.

Earth moves in a path around, or **revolves** around, the Sun. The Sun is a **star**, a huge ball of very hot, glowing gases. A **planet** is a large, nearly round object that revolves around a star. Earth is one of eight planets in our solar system. A **solar system** is a star and all the planets and other objects that revolve around that star.

Earth and the other planets revolve around the Sun. ▼

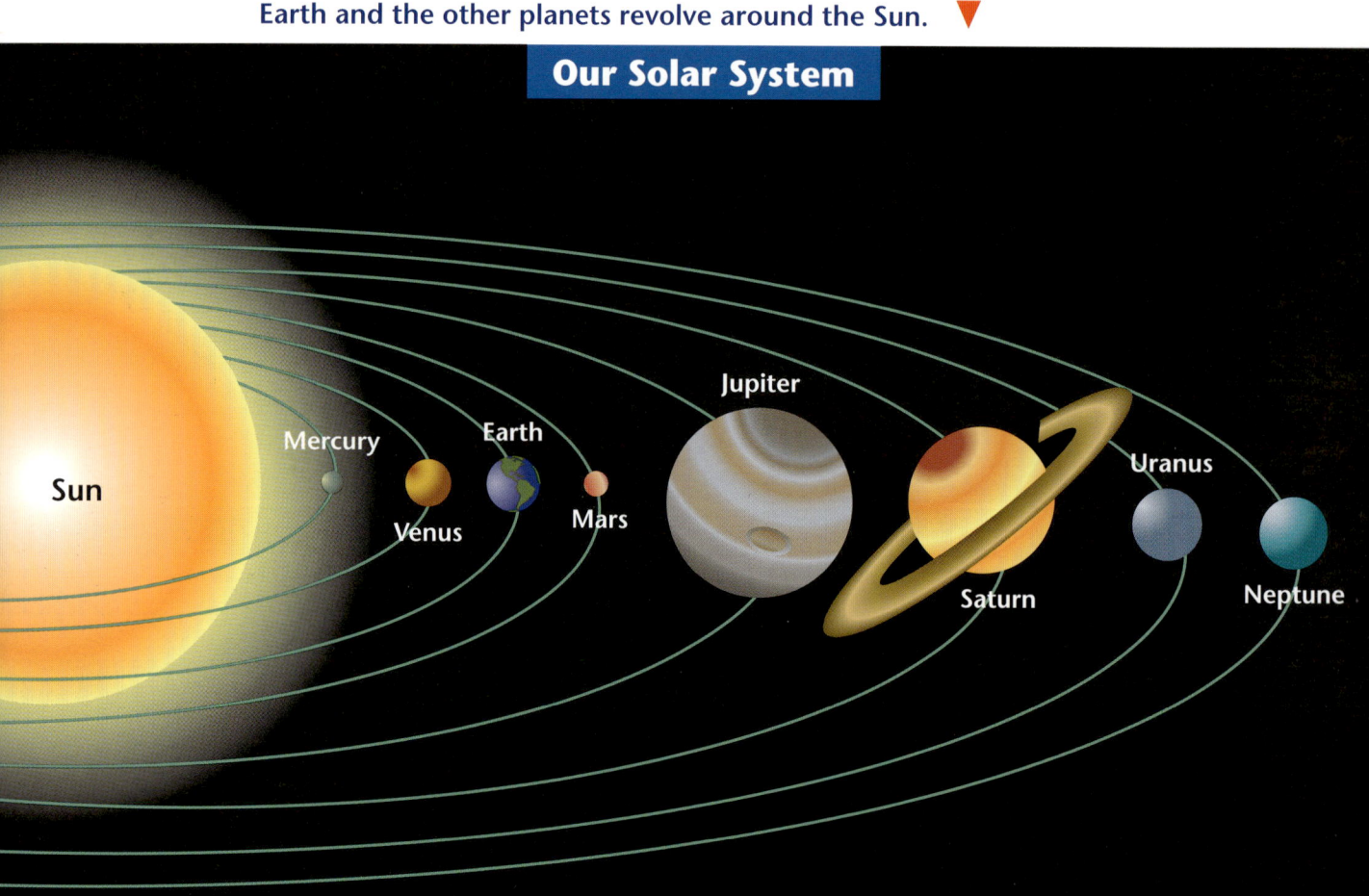

Our Solar System

This picture is not to scale.

The path a planet takes as it revolves around the Sun is called the planet's **orbit**. Each planet is a different distance from the Sun. So the orbits of the planets are different in size. The orbits are not perfect circles. They are oval shaped, or elliptical.

A year is the amount of time a planet takes to revolve around the Sun once. Earth's year is about 365 days long. Other planets take different lengths of time to revolve once.

Planets also spin, or **rotate**, on an axis. An **axis** is an imaginary line through the center of a planet. A day is the amount of time a planet takes to rotate once. Earth takes about 24 hours to rotate once. Other planets rotate at different speeds.

Each planet spins, or rotates, on its axis. ▼

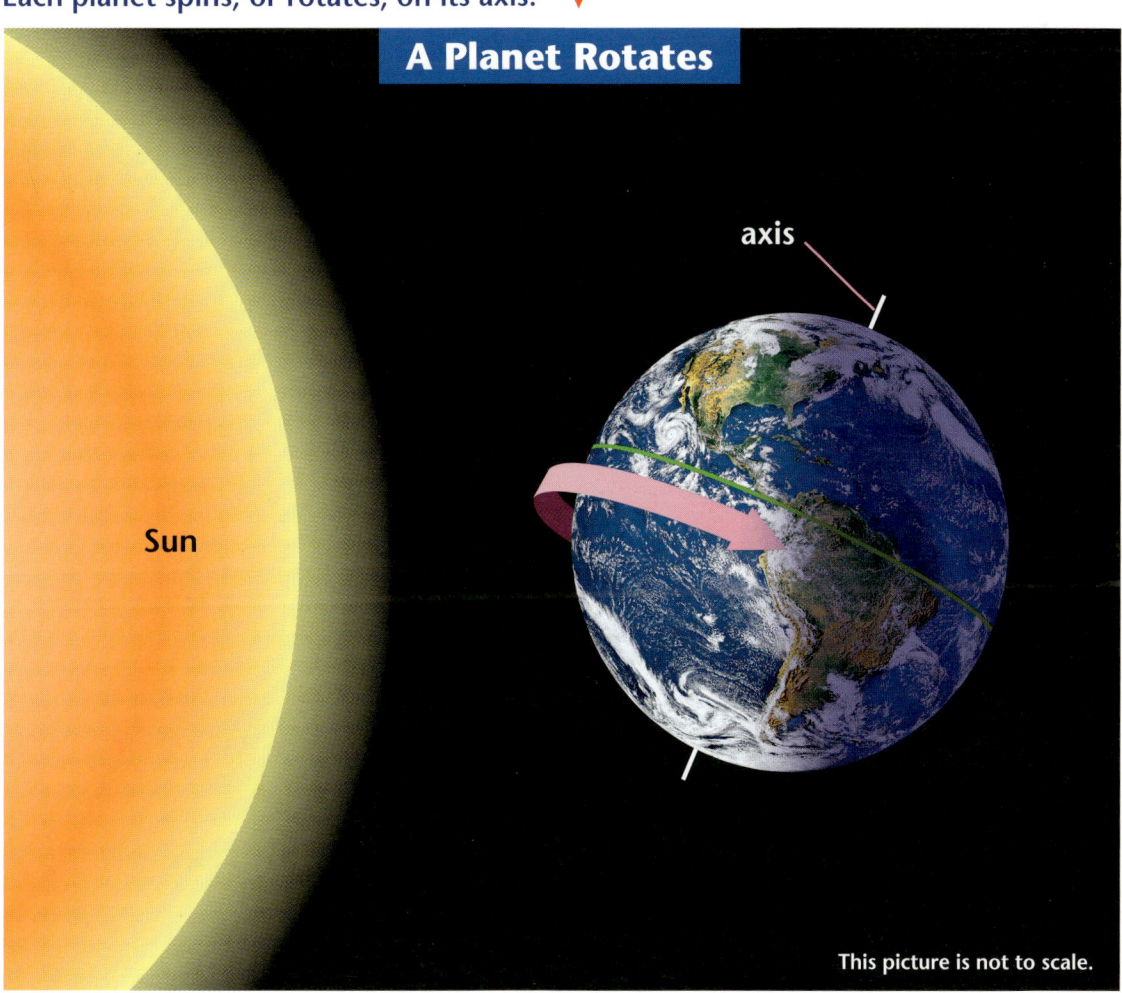

This picture is not to scale.

Gravity is the force that pulls all objects toward one another. Gravity between the Sun and the planets keeps the planets in their orbits.

Gravity is stronger between objects that have more mass. Mass is the amount of material in an object. The Sun has a huge mass. Gravity also is stronger between objects that are closer together.

Gravity acts everywhere. For example, gravity holds objects on Earth's surface. If you drop a pencil, it will fall to the floor. Gravity acts between Earth and the pencil. The pencil is pulled toward Earth.

Weight is a measurement of the force of gravity on an object. The pull of gravity at the surface of each planet is different. That is because each planet has a different mass and diameter. So the same object would have a different weight on different planets.

 What is a solar system?

If you weigh this much on Earth	Weight on Other Objects in Space			
^^	You would weigh about this much on			
^^	Mercury	Venus	Earth's Moon	Mars
60 pounds	23	54	10	23
100 pounds	38	90	17	38

▲ Your weight would be different on other objects in space. You would weigh less on an object with a weaker pull of gravity.

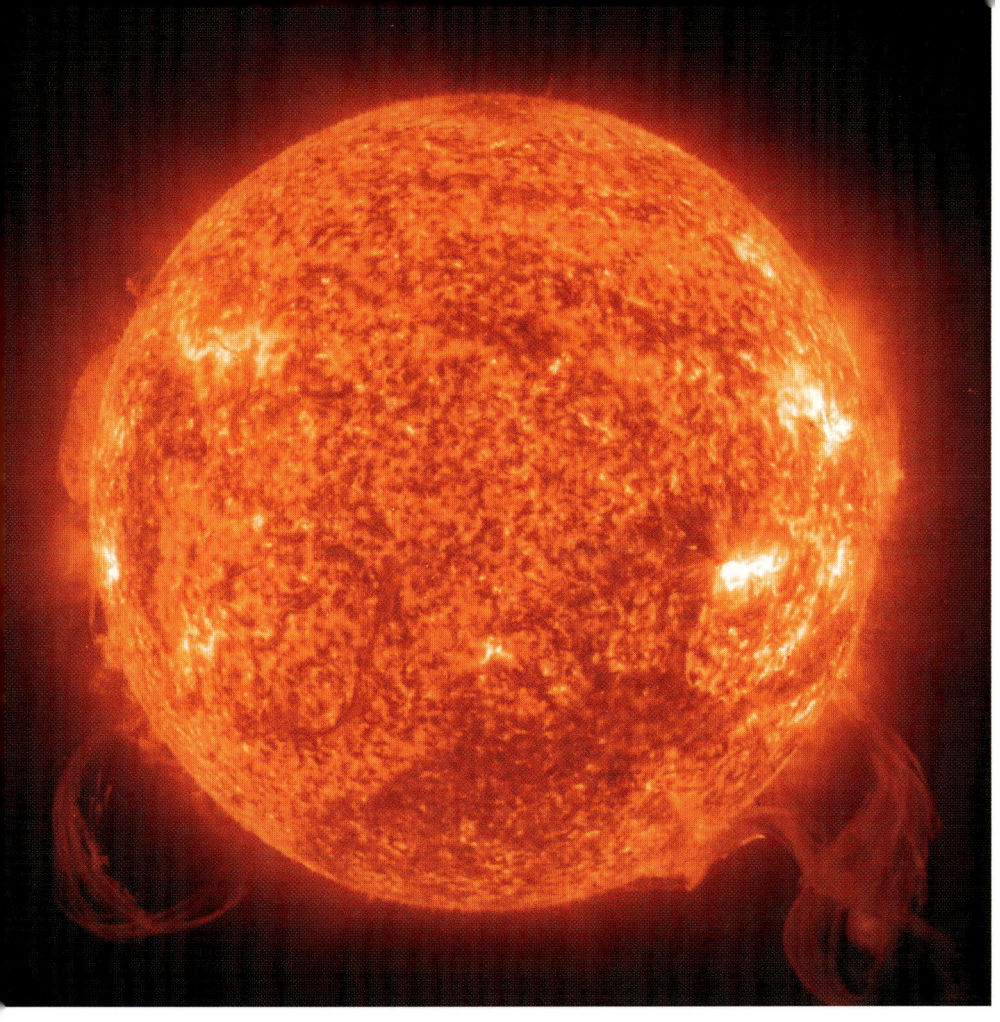

◀ The Sun is about 1,390,000 kilometers (about 864,000 miles) wide. If the Sun were an empty ball, about one million Earths would fit inside it.

The Sun

At the center of our solar system is the **Sun**. The Sun is the largest object in the solar system.

The Sun is the closest star to Earth. The Sun looks much larger to us than other stars do. But this is only because the Sun is so much closer to us than they are.

The Sun is made mostly of the gases hydrogen and helium. It gives off a huge amount of energy. Energy from the Sun moves out through space.

The Sun is the main source of energy for Earth. Most living things depend on light and heat from the Sun. Plants need sunlight to make their own food. Energy in that food passes to animals that eat plants. The Sun's energy also helps Earth stay at temperatures that support life.

✔ Why does the Sun look larger to us than other stars do?

The Planets

The largest objects in our solar system after the Sun are the eight planets. Earth is the third planet from the Sun. We can see some of the other planets from Earth. They look like bright dots of light in the sky. The planet Venus is not a star, but it is sometimes called the "morning star" or the "evening star." This is because Venus can be seen in the sky around sunrise or sunset.

Many planets have moons. A **moon** is a natural, rocky or icy object that revolves around a planet. Earth has one moon. The Moon can be seen sometimes during the day and sometimes at night. It is Earth's closest neighbor in space. The Moon stays in its orbit because of gravity between Earth and the Moon.

This picture is not to scale.

▲ One moon revolves around Earth.

▲ Mars is an inner planet. Its surface is solid and rocky.

Mercury, Venus, Earth, and Mars are the *inner planets*. They are closest to the Sun. These planets are small and dense. Their surfaces are solid and rocky. The inner planets have few or no moons.

Jupiter, Saturn, Uranus, and Neptune are the *outer planets*. They are farther from the Sun, so they are very cold. The outer planets are huge. They have rings and many moons. These planets are made mostly of gases. They are often called the gas giants.

The planets are very far apart. Even moving at a great speed, the space probe Voyager 2 took 12 years to fly from Earth to Neptune.

Saturn is an outer planet. It has rings and is made mostly of gases. ▼

The Inner Planets

MERCURY

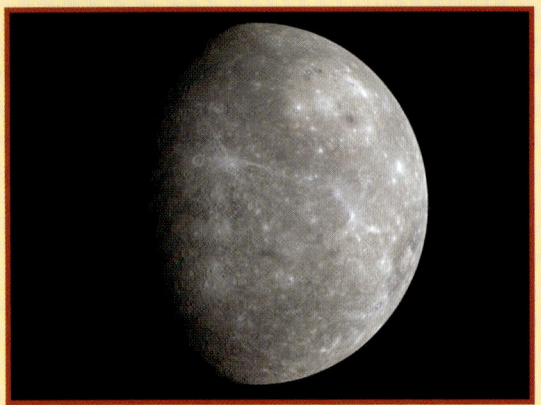

Mercury Facts
- Diameter: 4,879 km (3,032 mi)
- Surface: Rocky with craters, high cliffs, and smooth plains
- Moons: 0
- Length of Day: 58.6 Earth days
- Length of Year: 88 Earth days

Mercury is the smallest planet. It is less than half the size of Earth. Mercury is the closest planet to the Sun. Mercury has a very short year. The planet revolves around the Sun in only 88 Earth days.

Daytime on Mercury is hotter than Earth's hottest desert. At night, Mercury is colder than Antarctica. Mercury's rocky surface has bowl-shaped landforms called craters. The planet also has high cliffs and smooth plains.

VENUS

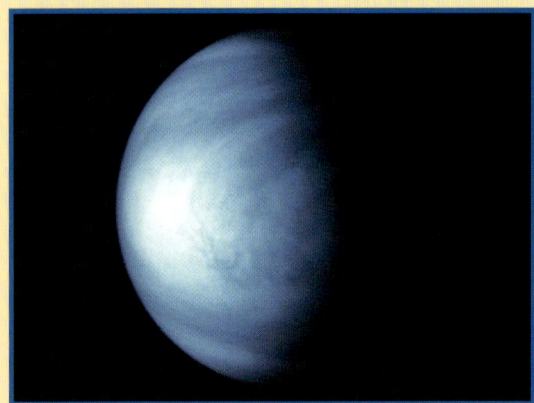

Venus Facts
- Diameter: 12,104 km (7,521 mi)
- Surface: Smooth plains, huge hills, volcanoes, and craters
- Moons: 0
- Length of Day: 243 Earth days
- Length of Year: 224.7 Earth days

Venus is about the same size as Earth. Venus rotates so slowly that a day on Venus is longer than a year on Venus!

Like many other planets, Venus has an atmosphere. An atmosphere is a layer of gases around a space object. Venus's atmosphere is mostly carbon dioxide. This gas traps the Sun's heat very well. Because of this, Venus is the hottest planet. Thick clouds cover Venus.

EARTH

Earth Facts
- Diameter: 12,756 km (7,926 mi)
- Surface: Rocky with plains, mountains, volcanoes, and craters; close to ¾ covered with water
- Moons: 1
- Length of Day: 24 hours
- Length of Year: 365.25 days

Earth is the only planet known to support life. Our "blue planet" has a lot of water. Some of that water is in liquid form. Living things need water. Earth has an atmosphere with gases, such as oxygen, that living things need. Earth's atmosphere and distance from the Sun help Earth stay at temperatures that support life.

Earth's moon is rocky and dusty. It has plains, mountains, valleys, and many craters. The Moon has almost no atmosphere.

MARS

Mars Facts
- Diameter: 6,792 km (4,220 mi)
- Surface: Desert with highlands, lowlands, craters, and volcanoes
- Moons: 2
- Length of Day: 24 Earth hours, 37 minutes
- Length of Year: 687 Earth days

Mars is called the "red planet" because of its reddish color. It is smaller than Venus or Earth. The land on Mars is something like Earth's deserts. Mars has hills, valleys, and many craters. It also has many old volcanoes. Frozen water has been found on Mars.

Mars has a thin atmosphere. The average surface temperature on Mars is about -63° Celsius (about -81° Fahrenheit).

The Outer Planets

JUPITER

Jupiter Facts
- Diameter: 142,984 km (88,846 mi)
- Surface: Gases
- Moons: At least 63
- Length of Day: 9 Earth hours, 55 minutes
- Length of Year: 11.9 Earth years

Jupiter is our solar system's largest planet. It is so big that all the other planets together could fit inside it. Jupiter has an atmosphere that is made mostly of the gases hydrogen and helium. So do the other gas giants.

Jupiter has bands of clouds. The planet's Great Red Spot is a giant storm that has lasted for more than 300 years. Jupiter has at least three rings.

SATURN

Saturn Facts
- Diameter: 120,536 km (74,898 mi)
- Surface: Gases
- Moons: At least 60
- Length of Day: 10 Earth hours, 39 minutes
- Length of Year: 29.5 Earth years

Saturn is the second-largest planet in our solar system. Its bright rings are made of pieces of ice and rock. Saturn is lighter than water. If you could drop Saturn into a huge bathtub, it would float!

Saturn's largest moon is called Titan. Scientists think Titan's atmosphere is much like the atmosphere Earth had long ago.

URANUS

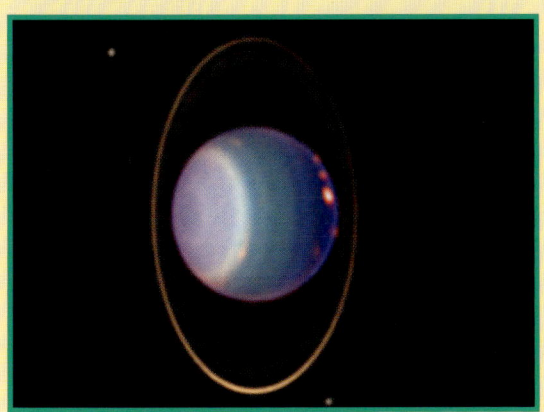

Uranus Facts
- Diameter: 51,118 km (31,763 mi)
- Surface: Gases
- Moons: At least 27
- Length of Day: 17 Earth hours, 14 minutes
- Length of Year: 84 Earth years

Uranus was discovered in 1781. It was the first planet to be discovered with a telescope. Uranus's axis is very tilted. So the planet rotates almost on its side. Methane gas in Uranus's atmosphere makes the planet look bluish green.

Uranus has at least 13 rings. The space probe Voyager 2 helped us discover 11 of Uranus's moons. It flew by Uranus in 1986.

NEPTUNE

Neptune Facts
- Diameter: 49,528 km (30,775 mi)
- Surface: Gases
- Moons: At least 13
- Length of Day: 16 Earth hours
- Length of Year: 164.8 Earth years

Neptune is the farthest planet from the Sun. It is sometimes called Uranus's twin planet. It is about the same size as Uranus and is made of similar gases.

Strong winds blow on all the gas giant planets. However, Neptune's winds are the fastest. They can blow at speeds of about 2,000 kilometers (about 1,200 miles) per hour. Neptune has at least five rings. It also has at least 13 moons.

 Tell about two ways that the inner planets and the outer planets are different.

Other Objects in the Solar System

Our solar system includes many objects other than the planets and their moons. These other objects also revolve around the Sun.

Dwarf planets have some features of planets. They are nearly round and revolve around the Sun. But they are smaller than planets. Pluto, Ceres, and Eris are three dwarf planets.

At one time, Pluto was called a planet. But it was somewhat different from other planets. With new, stronger telescopes, scientists found other faraway objects like Pluto. Scientists now group these objects together as dwarf planets.

Asteroids are space objects made of rock, metal, or a mixture of the two. Most asteroids are less than 1 kilometer (about 0.62 miles) wide. But some are much larger. Most asteroids are found in an area between Mars and Jupiter. This area is called the *asteroid belt*.

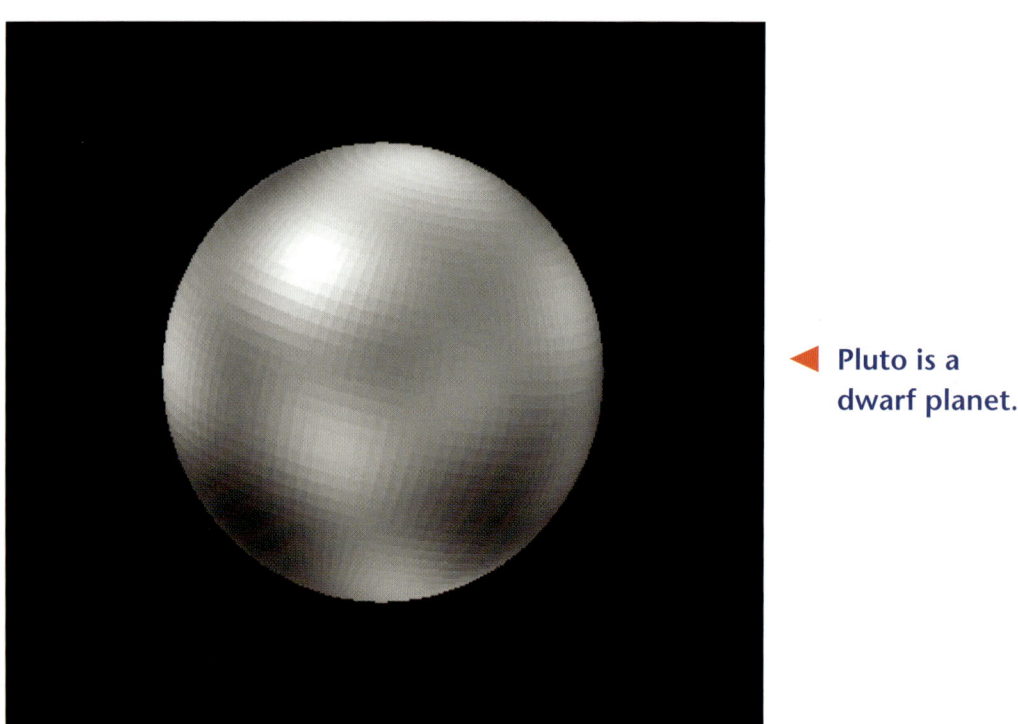

◀ Pluto is a dwarf planet.

▲ Halley's comet passes close to Earth about every 76 years.

▲ A meteor is a streak of light made when a meteoroid enters Earth's atmosphere.

Comets are small space objects made of frozen gases, ice, rock, and dust. Comets revolve around the Sun in long, elliptical orbits. If a comet moves near the Sun, some of its ice changes to gas. A tail of gas and dust forms behind the comet.

Meteoroids are small pieces of rock, metal, or both. Some were once pieces of asteroids or comets. Meteoroids sometimes enter Earth's atmosphere. As a meteoroid moves through the air, it heats up. It may make a streak of light in the sky. This brightly burning meteoroid is called a **meteor** or "shooting star." A meteor may not burn up completely. A piece that hits Earth's surface is called a *meteorite*. Large meteorites can form craters.

 How are asteroids different from comets?

REFLECT ON READING
You previewed pictures, captions, and other book features before reading. Tell how one book feature helped you better understand the solar system or an object in it.

APPLY SCIENCE CONCEPTS
Think about how important the Sun is to the solar system. What might be different about the planets without the Sun? Tell about your ideas.

Build Reading Skills

Main Idea and Details

The **main idea** of a paragraph or part of a book is the most important point. **Details** give more information about the main idea.

As you read page 20, look for the main idea about galaxies.

TIPS

The topic sentence tells the main idea of a paragraph. It is often the first sentence in the paragraph. To find the main idea, ask, "What is this paragraph mostly about?"

Details may answer Who, What, When, Where, Why, and How questions about the main idea. Details can be

- examples
- descriptions
- reasons
- other facts

A concept web can help you keep track of the main idea and details.

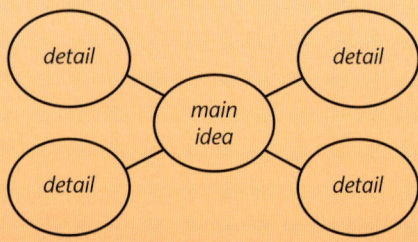

What Is Beyond Our Solar System?

MAKE A CONNECTION
Look at this picture of a faraway part of space. What do you see? What do you think is beyond our solar system?

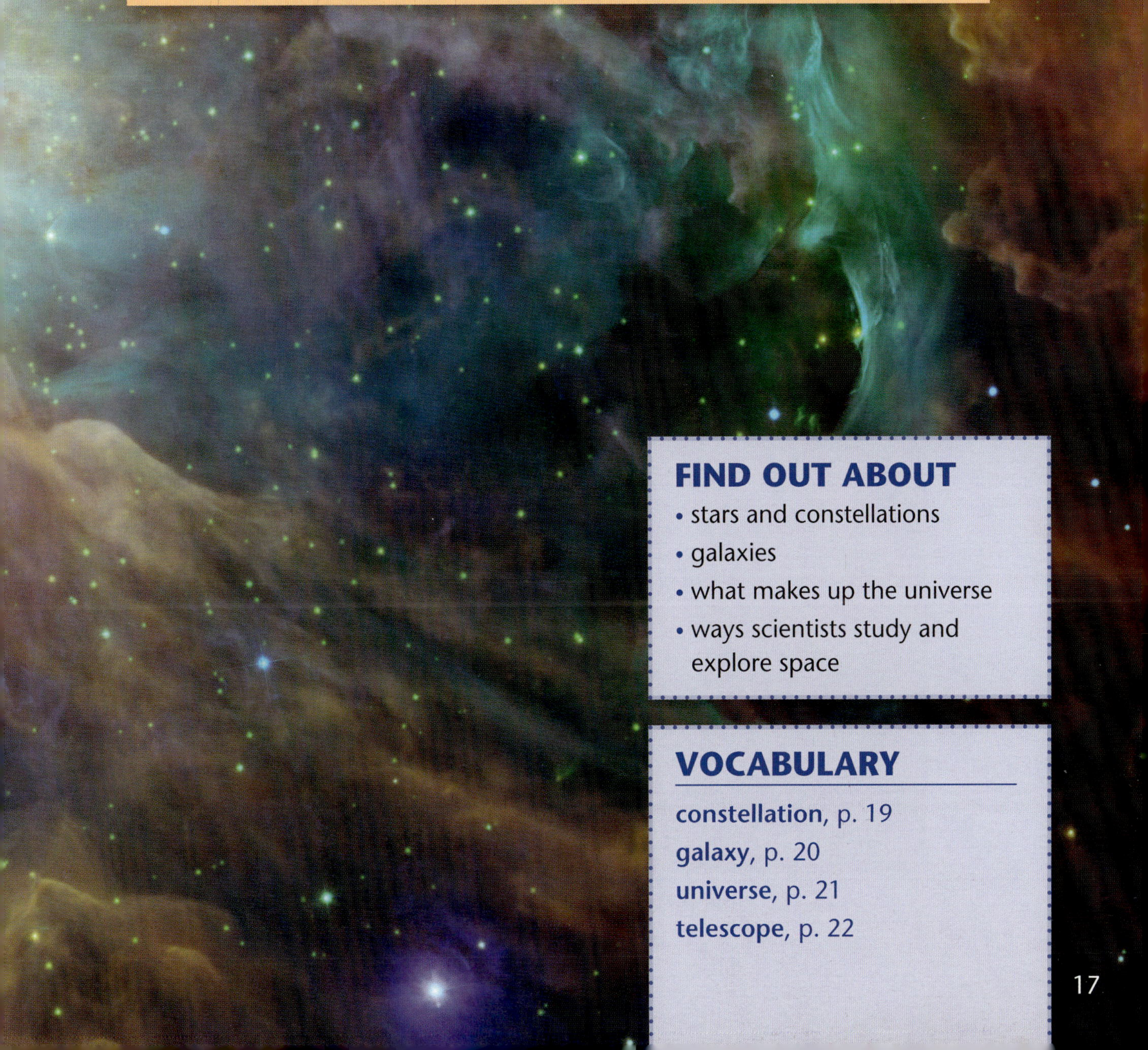

FIND OUT ABOUT
- stars and constellations
- galaxies
- what makes up the universe
- ways scientists study and explore space

VOCABULARY
constellation, p. 19
galaxy, p. 20
universe, p. 21
telescope, p. 22

Other Stars

The Sun is just one of a huge number of stars in space. Other stars look like dots of light to us. This is because they are so far away. Most stars are separated by huge distances. Two stars that look close together in the sky may actually be very far apart.

Stars can be grouped, or classified, by size or by brightness. Our Sun is a medium-sized star. It has an average brightness.

Stars also can be classified by their color and temperature. Hotter stars are blue or white in color. Cooler stars are orange or red.

▲ Stars can be different in size, brightness, color, and temperature.

▶ The stars that make up the Big Dipper look as if they are close together. They are actually very far apart.

A **constellation** is a group of stars that make a pattern in the sky. The pattern of stars in a constellation always looks the same.

Many constellations are named after animals or characters in stories from long ago. The constellation Orion is named after a hunter in Greek myths. The Big Dipper is a part of the constellation Ursa Major, or Great Bear. The Little Dipper is part of Ursa Minor, or Little Bear.

The constellations appear to move slowly across the sky during the night. This happens because Earth rotates. Some constellations can be seen only during certain seasons. This is because Earth changes position in space as it revolves around the Sun.

✔ Name some ways that stars can be grouped, or classified.

Galaxies

A **galaxy** is a huge collection of stars, dust, and gas held together by gravity. A galaxy may have hundreds of billions of stars. Billions of galaxies are in space. Our solar system is a part of the Milky Way galaxy.

The distances between most galaxies are huge. Galaxies are not spread evenly throughout space. Instead, they are often found in groups, or clusters.

Galaxies are grouped, or classified, by shape. There are three main kinds.

- *Spiral* galaxies are shaped like pinwheels, with "arms" that swirl out. The Milky Way is a spiral galaxy.
- *Elliptical* galaxies are oval shaped.
- *Irregular* galaxies have no regular shape.

✓ Tell about the three main kinds of galaxies.

▲ Galaxy M51, also called the Whirlpool galaxy, is a spiral galaxy.

Galaxy NGC 4449 is an irregular galaxy. ▶

20

▲ This picture is from the Hubble Space Telescope. It shows hundreds of galaxies in just one part of the sky.

The Universe

The **universe** is all of space and everything in it. All the planets, stars, galaxies, and other objects in space are part of the universe. So is all the empty space between them.

The huge size of the universe is hard to imagine. Most of the galaxies in the universe are very far apart. Scientists have found that most galaxies are moving even farther apart from one another. So scientists think that the universe is getting larger, or expanding. New technology will help scientists learn more about the universe and how it is changing.

 What does the universe include?

Studying and Exploring Space

A **telescope** is a tool that helps us see faraway objects. The telescope was invented in the early 1600s. Galileo Galilei of Italy was the first scientist to use a telescope to look at objects in space.

Some telescopes have rounded pieces of glass called lenses. The lenses magnify objects. Objects such as the Moon, planets, and stars look closer and larger when seen through the lenses.

Telescopes help scientists study space. We can see many more space objects using a telescope than we can see with our eyes alone.

Scientists use telescopes that are on Earth and in space. Large telescopes on Earth are usually put on high mountains. Space telescopes, such as the Hubble Space Telescope, are put into orbit above Earth's atmosphere.

◀ A telescope makes faraway objects look closer and larger.

▲ Astronauts have used spacecraft to visit and study the Moon.

▲ Scientists live and work in the *International Space Station*.

Scientists also study space with spacecraft and space probes. In the 1950s and 1960s, people worked to build the first rockets able to go into space. A United States spacecraft landed on the Moon in 1969. Astronaut Neil Armstrong became the first person to walk on the Moon.

Today, spacecraft called space shuttles carry astronauts into space. They sometimes go to the *International Space Station*. It is a science lab that is in orbit around Earth.

Space probes are spacecraft that carry cameras and tools, but not people. They have been sent to planets, moons, asteroids, and comets. Space probes send information back to Earth.

 What are some ways scientists study space?

REFLECT ON READING
Make a concept web like the one on page 16. Use it to keep track of information about galaxies. Put the main idea in the middle. Add details, such as examples or other facts.

APPLY SCIENCE CONCEPTS
Why do you think scientists use space probes to study our solar system? Why might using probes be better than using telescopes or sending people into space? Write your ideas in your science notebook.

Glossary

asteroid (AS-tuh-roid) a small object made of rock, metal, or a mixture of the two that revolves around the Sun **(14)**

axis (AK-sis) an imaginary line that goes through the center of an object, around which the object spins, or rotates **(5)**

comet (KOM-it) a frozen mass of gases, ice, rock, and dust that revolves around the Sun **(15)**

constellation (kon-stuh-LAY-shuhn) a group of stars that make a pattern in the sky **(19)**

dwarf planet (DWORF PLAN-it) a small, nearly round object that revolves around the Sun and has some features of a planet **(14)**

galaxy (GAL-uhk-see) a huge collection of stars, dust, and gas held together by gravity **(20)**

meteor (MEE-tee-uhr) a brightly burning meteoroid, a chunk of rock or metal from space, that is falling through Earth's atmosphere **(15)**

moon (MOON) a natural, rocky or icy object that revolves around a planet **(8)**

orbit (OR-bit) the path an object takes as it revolves around another object in space **(5)**

planet (PLAN-it) a large, nearly round object that revolves around a star **(4)**

revolve (rih-VAHLV) to move in a path, or orbit, around another object **(4)**

rotate (RO-tayt) to spin on an axis **(5)**

solar system (SOH-luhr SIS-tuhm) a star and the planets, their moons, and other objects that revolve around that star **(4)**

star (STAHR) a huge ball of very hot, glowing gases in space that gives off energy **(4)**

Sun (SUHN) the star that is at the center of our solar system **(7)**

telescope (TEL-i-skohp) a tool that makes faraway objects look closer and larger by using lenses, mirrors, or other devices **(22)**

universe (YOO-nuh-vurs) all of space and everything in it **(21)**